ASTRONOMY UNVEILED

A Journey Through the Cosmos

Philipp Frühwirth

CONTENTS

INTRODUCTION TO ASTRONOMY

Humans have looked up at the stars since the beginning of civilization. The sheer vastness and complexity of the universe have piqued our curiosity and imagination for thousands of years. Astronomy, the scientific study of celestial objects and their properties and behavior, has enabled us to investigate the universe in a more systematic and rigorous way.

The origins of astronomy can be traced back to ancient civilizations such as the Babylonians, Greeks, and Egyptians who observed and recorded the movements of celestial objects. However, it was not until the development of the telescope and other advanced technologies that our understanding of the universe expanded significantly.

The study of astronomy is a multidisciplinary field that encompasses physics, chemistry, mathematics, and other natural sciences. It involves the observation and analysis of various celestial phenomena, including stars, planets, asteroids, comets, galaxies, and cosmic rays. Through this study, we hope to uncover the nature of the universe, its origin, and its evolution.

One of the major contributions of astronomy to our understanding of the universe is the discovery that it is ever-expanding. In 1929, astronomer Edwin Hubble observed that galaxies were moving away from each other, leading to the realization that the universe is expanding from a single point in time, known as the Big Bang. This discovery has become a cornerstone of modern cosmology and is supported by a vast body of evidence gathered through astronomical observations.

Another important aspect of astronomy is the search for life beyond our planet. Given the vastness of the universe and the fact that Earth is not the only planet in the universe with conditions suitable for life, astronomers are interested in finding life on other planets. They are looking at the possibility of exoplanets, planets outside our solar system that could have suitable conditions for life to emerge.

Astronomy also plays a role in the development of technology. Many of the technologies used in astronomy, including telescopes, detectors, and imaging systems, have found applications in other fields. For example, medical imaging techniques such as X-rays, magnetic resonance imaging (MRI), and positron emission tomography (PET) are based on similar principles employed in astronomical observations.

In conclusion, astronomy is a fascinating field that enables us to explore the universe and deepen our understanding of our place in it. Over the years, we have made remarkable progress in uncovering the mysteries of the universe, and there is much more to learn. Astronomy is a continually evolving field, and new discoveries and breakthroughs are awaited with anticipation.

HISTORICAL DEVELOPMENTS IN ASTRONOMY

The study of astronomy dates back to ancient times, when people looked up at the night sky and wondered about the stars, planets, and other celestial bodies. Over the centuries, many cultures developed myths and legends about the heavens, but it wasn't until the ancient Greeks began to develop the scientific method that astronomy began to take on a more rigorous approach.

One of the most important figures in early astronomy was the Greek philosopher Aristotle, who lived in the 4th century BCE. He believed in a geocentric universe, with Earth at the center and everything else revolving around it. This model was widely accepted for centuries, until the Polish astronomer Copernicus proposed a heliocentric model in the 16th century.

The 16th and 17th centuries were a time of rapid progress in astronomy, thanks in part to the invention of the telescope. In 1609, Galileo turned his telescope to the skies and observed the moons of Jupiter, proving the heliocentric model and demonstrating the potential of this new tool.

Isaac Newton, another towering figure in the history of science, revolutionized astronomy in the 17th century with his laws of motion and universal gravitation. These laws allowed astronomers to predict the motions of celestial bodies and paved the way for the discovery of new planets and other objects.

The 18th and 19th centuries brought new advances in the study of astronomy, including the discovery of Uranus, Neptune, and Pluto, as well as the development of spectroscopy, which allowed astronomers to analyze the chemical composition of stars and other bodies.

In the 20th century, astronomy continued to make rapid progress, with breakthroughs including the discovery of black holes, the development of radio astronomy, and the launch of the first artificial satellites. The Hubble Space Telescope, launched in 1990, has provided stunning images of the universe and allowed astronomers to study the most distant objects ever observed.

Today, astronomy continues to be a dynamic and exciting field, with new discoveries and breakthroughs happening all the time. From searching for exoplanets to studying the nature of dark matter and dark energy, there is still so much to learn about the universe we inhabit.

THE SOLAR SYSTEM

The solar system refers to the collection of astronomical objects that orbit around the Sun. The Sun is the largest object in the solar system, followed by the eight planets, dwarf planets, asteroids, and comets. The solar system formed around 4.6 billion years ago from a cloud of gas and dust, and has since remained relatively stable.

The eight planets in the solar system are classified into two main groups: the inner or terrestrial planets (Mercury, Venus, Earth, and Mars) and the outer or gas giant planets (Jupiter, Saturn, Uranus, and Neptune). The terrestrial planets are generally small and rocky, while the gas giant planets are much larger and composed mainly of hydrogen and helium.

Mercury is the smallest planet in the solar system and closest to the Sun, with a rocky surface similar to that of the Moon. Venus is the second planet from the Sun and is known for its thick and toxic atmosphere, making it the hottest planet in the solar system. Earth, meanwhile, is the third planet from the Sun and is the only planet known to harbor life. Mars is the fourth planet from the Sun and has a thin atmosphere, but extensive evidence of water in the past suggests it may have once been able to support life.

Jupiter is the largest planet in the solar system and is known for its prominent colored bands and the Great Red Spot, a giant storm that has been raging for over 300 years. Saturn is the sixth planet from the Sun and is perhaps best known for its spectacular rings. Uranus and Neptune are the outermost gas giant planets, and are often referred to as ice giants due to their composition of water, ammonia, and methane.

The solar system also contains several dwarf planets, the most famous of which is Pluto. Discovered in 1930 and considered the ninth planet in the solar system for almost 70 years, Pluto was reclassified as a dwarf planet in 2006.

In addition to the planets and dwarf planets, the solar system is home to numerous asteroids, which are small rocky objects that orbit the Sun. Some asteroids have even been visited by spacecraft, such as the Dawn mission to the asteroid belt. The solar system also contains comets, which are icy objects that typically have highly elliptical orbits and produce spectacular tails when they approach the Sun.

Studying the solar system and its objects is important for understanding the origins of the universe and the potential for life beyond Earth. Thanks to numerous space missions, our knowledge of the solar system continues to expand and evolve.

THE SUN: OUR CLOSEST STAR

The sun is the brightest and largest object in our solar system. It is a star, and it is the center of our solar system around which all the other planets revolve. It has been an object of fascination for humans for thousands of years, and it is still a subject of intensive scientific research.

The sun has a diameter of around 1.39 million km, which is about 109 times that of the Earth. Its mass is 330,000 times that of the Earth, and it accounts for more than 99.8% of the total mass of the solar system. The sun is a middle-aged star, currently in the phase in which it fuses hydrogen atoms into helium atoms in its core. This process is called nuclear fusion, and it releases a tremendous amount of energy in the form of light and heat, which provide the energy for life on Earth.

The sun's surface, which is visible to us, is covered by convection cells, which produce the granular appearance. The sunspots, which are darker regions on the sun's surface, are areas of intense magnetic activity. They are associated with the release of energy in the form of solar flares and coronal mass ejections, which can impact our planet's magnetic field and cause auroras in the polar regions.

The sun's atmosphere is composed of several layers, including the photosphere, the chromosphere, and the corona. The photosphere is the layer that emits most of the visible light, and it is the layer that we see when we observe the sun. The chromosphere is the layer above the photosphere and is hotter than the photosphere. The corona is the outermost layer, which extends millions of kilometers into space and is visible during total solar eclipses.

The sun is a dynamic and complex object, and scientists are

still discovering new things about it. With the help of modern telescopes and space probes, we can study the sun's magnetic field, its interior structure, and its influence on the solar system. Understanding the sun's behavior is crucial for predicting and mitigating the effects of space weather on our technology and infrastructure on Earth.

In conclusion, the sun is a remarkable object that has played a fundamental role in the history of our planet and continues to have a profound impact on our lives. As we learn more about the sun, we gain a deeper appreciation of its complexity and beauty.

THE MOON

The moon is the Earth's only natural satellite and the fifth largest moon in the solar system. It is one of the most recognizable objects in the night sky and has been a subject of human fascination for centuries. The moon's proximity to Earth makes it an ideal target for space exploration, and it has been visited by many unmanned and manned missions.

The moon was formed around 4.5 billion years ago as a result of a giant impact between Earth and a Mars-sized object. This collision ejected a large amount of debris into space that eventually coalesced to form the moon. The moon's relatively large size compared to Earth's other natural satellites is a result of its formation process.

The moon's surface is covered with an array of features including mountains, craters, valleys, and plains. These features were formed by a combination of internal and external geological processes that took place over millions of years. The moon's lack of an atmosphere has preserved many of these features, providing scientists with a valuable window into the early history of the solar system.

The moon's surface is heavily cratered, which suggests that it has undergone a long history of impact events. The craters on the moon range in size from tiny pits to huge, multi-ringed basins that are hundreds of kilometers across. Some of the largest craters, known as impact basins, are thought to have been created by collisions with objects the size of small planets.

The moon's lack of an atmosphere means that it is exposed to space weather, including solar winds, cosmic rays, and meteoroid impacts. These factors have contributed to the moon's bleak and barren landscape. However, the moon is still an important target for scientific exploration, as it provides clues to the earliest history of the solar system and the formation of the Earth.

In recent years, there has been renewed interest in returning humans to the moon. The goal of these missions is to establish a sustainable human presence on the moon that will enable scientific research, resource utilization, and eventual exploration of other targets in the solar system. These missions will build on the achievements of previous lunar exploration efforts, including the Apollo program that sent astronauts to the moon in the 1960s and 1970s.

Overall, the moon is a fascinating object that continues to capture the interest of scientists and the public alike. Its unique properties and proximity to Earth make it a valuable target for scientific exploration and a potential destination for future human exploration.

PLANETS IN THE
HABITABLE ZONE

With the advancements in modern technology and space exploration, scientists have been able to identify many planets orbiting other stars in our galaxy. Among these, the planets that lie in the habitable zone of their stars have garnered a lot of attention in the scientific community. The habitable zone refers to the region around a star where the temperatures are just right to allow liquid water to exist on the surface of a planet. This is an important criterion for the existence of life as we know it.

The habitable zone depends on several factors such as the size, mass, and temperature of the star, as well as the planet's distance from it. M-dwarf stars, which are the most common stars in our galaxy, have smaller habitable zones than sun-like stars. Planets that are closer to the star than the inner edge of the habitable zone would have runaway greenhouse effects, like Venus in our solar system, while planets further out would be too cold to sustain liquid water.

Some of the most promising exoplanets in the habitable zone identified so far include Proxima b, which orbits the nearest star to our solar system, Proxima Centauri. Another exciting prospect is TRAPPIST-1, a system of seven Earth-sized planets orbiting a cool M-dwarf star. The discovery of these exoplanets has opened up new avenues for research in astrobiology and the search for life beyond Earth.

The NASA Kepler mission has been instrumental in identifying the first exoplanets in the habitable zone. It observed a patch of the sky in the constellation Cygnus, discovering thousands of exoplanet candidates. Kepler-22b, Kepler-438b, Kepler-62e,

Kepler-62f, and Kepler-186f are some of the exoplanets discovered by the Kepler mission that lie in the habitable zone of their respective stars.

Scientists are currently studying the atmospheres of these exoplanets to look for signs of life. Among the various techniques used for this purpose, the transit spectroscopy method involves measuring the absorption of light by the planet's atmosphere as it transits across the face of its star. This method has been successful in detecting water vapor and other gases in the atmospheres of some exoplanets.

The discovery of exoplanets in the habitable zone has sparked the imagination of the general public and has helped to stimulate interest in astronomy and space exploration. It holds the promise of discovering extraterrestrial life and expanding our understanding of the universe and our place in it.

ASTEROIDS AND COMETS

Asteroids and comets are small, rocky objects that exist within our solar system. These objects can be found orbiting the Sun, and they can provide valuable insights into the formation and evolution of our solar system.

Asteroids are small, rocky objects that orbit the Sun. Most asteroids can be found within the asteroid belt, which exists between Mars and Jupiter. Asteroids can vary in size from just a few meters in diameter to several hundred kilometers in diameter. While some asteroids may be relatively smooth and featureless, others may be covered in craters and other features resulting from impacts with other asteroids or comets.

Comets, on the other hand, are icy bodies that are similar in composition to asteroids but with a much different appearance. Comets are often described as "dirty snowballs" because they are made up primarily of water, ice, and dust. As comets approach the Sun, the heat causes the ice to vaporize, creating a bright coma and a tail of gas and dust pointing away from the Sun.

One of the most famous comets is Halley's comet, which visits our inner solar system every 76 years. When it is at its closest approach to the Sun, Halley's comet can be seen from Earth as a bright streak in the sky.

Asteroids and comets are both of great interest to astronomers because they can provide insight into the origins of the solar system. By studying the composition and distribution of asteroids and comets, scientists can learn more about the processes that led to the formation of the planets and other bodies within our solar system.

Additionally, asteroids and comets can pose a potential threat to

Earth. While the chances of a catastrophic impact with a large asteroid or comet are relatively small, the consequences could be severe. Consequently, astronomers continue to study these objects in an effort to better understand their orbits and potential impact risks.

In recent years, NASA has sent spacecraft to study both asteroids and comets up close. These missions, such as the Rosetta spacecraft's rendezvous with comet 67P/Churyumov-Gerasimenko and the Dawn spacecraft's exploration of the asteroid Vesta and dwarf planet Ceres, have provided valuable insights into the composition and structure of these bodies.

Overall, asteroids and comets are fascinating objects that have captured the attention of astronomers and the general public alike. Despite being relatively small and often overlooked, these objects play an important role in the history and evolution of our solar system.

TELESCOPES AND OBSERVATORIES

Telescopes and observatories are the backbone of the field of astronomy. They allow us to see and study the universe beyond the limits of our naked eye. Telescopes come in a variety of shapes and sizes, and each type is designed to capture a specific type of light. Some telescopes observe visible light, while others observe infrared, ultraviolet, or other forms of electromagnetic radiation. Observatories, on the other hand, are facilities where astronomers conduct their observations and gather the necessary data.

There are two main types of telescopes: refracting and reflecting telescopes. Refracting telescopes use lenses to bend and focus light, while reflecting telescopes use mirrors. Reflecting telescopes are cheaper to produce and can produce high-quality images with fewer aberrations, so they have become more popular in recent years. The largest reflecting telescope in the world is the Gran Telescopio Canarias in the Canary Islands, which has a mirror that is almost 11 meters in diameter.

There are also telescopes that observe radio waves and microwaves, such as the Atacama Large Millimeter Array (ALMA) in Chile. These telescopes are used to study the cold, molecular gas that makes up much of the universe, as well as to search for signs of extraterrestrial life.

Observatories are equipped with state-of-the-art instruments and facilities to help astronomers gather data in a more efficient manner. One example is the Keck Observatory, located atop Hawaii's Mauna Kea, which houses two of the world's largest telescopes. The observatory's Adaptive Optics System helps to reduce the blurring effects of the Earth's atmosphere, providing incredibly high-resolution images.

The Hubble Space Telescope, operated by NASA and the European Space Agency (ESA), is another example of an observatory. Unlike ground-based telescopes, Hubble is in orbit around the Earth, which allows it to capture images of objects and regions of space that are not visible from the ground. Hubble has made numerous discoveries and is regarded as one of the most important scientific instruments ever built.

In conclusion, telescopes and observatories are crucial tools for astronomers to observe and study the universe. With innovative advancements and a continuous search for new discoveries, we can expect numerous breakthroughs and progressions in the field of astronomy in the coming years.

THE HUBBLE SPACE TELESCOPE

Since its launch in 1990, the Hubble Space Telescope has revolutionized our understanding of the universe. It is one of the most important and successful scientific instruments ever created, enabling astronomers to observe the universe in a way that was previously impossible.

The Hubble Telescope was named after American astronomer Edwin Hubble, who is known for discovering that the universe is expanding. It was built as a joint project between NASA and the European Space Agency (ESA) and is operated by the Space Telescope Science Institute (STScI) in Baltimore, Maryland.

The telescope is located in low Earth orbit at an altitude of approximately 340 miles (550 km). This altitude provides excellent observing conditions, as the telescope is above the Earth's atmosphere, which can distort and blur images of space objects.

The Hubble Telescope has a primary mirror that is 2.4 meters (7.9 feet) in diameter, which collects and reflects light from deep space objects like galaxies, stars, and nebulas onto its various instruments. These instruments measure the brightness, temperature, and chemical composition of these objects, which helps astronomers to learn more about their nature and origins.

One of the most significant achievements of the Hubble Telescope has been its ability to observe and photograph distant galaxies. It has revealed that the universe is filled with billions of galaxies, each containing billions of stars. These observations have led to new theories about the formation and evolution of galaxies, as well as the age and size of the universe.

The telescope has also provided us with stunning images of

planets, moons, and other objects in our Solar System, including Jupiter's Great Red Spot and the rings of Saturn. These images have helped us to study and understand these objects in more detail.

Over its three decades of operation, the Hubble Telescope has undergone five servicing missions, which have repaired and upgraded its instruments, and extended its life expectancy. It has contributed greatly to our understanding of the universe and has inspired a generation of astronomers and space enthusiasts.

In summary, the Hubble Space Telescope is a truly remarkable scientific instrument that has allowed us to make groundbreaking discoveries about the universe. It has provided us with stunning images and important data that have expanded our knowledge of the cosmos, and it will continue to be an important tool for astronomy for many years to come.

THE MILKY WAY GALAXY

The Milky Way is a barred spiral galaxy that thousands of scientific studies have been conducted on. It is believed to be around 13.6 billion years old and has a diameter of approximately 100,000 light-years. The Milky Way is home to over 100 billion stars, and estimates suggest it contains upwards of 100 billion planets.

The structure of the Milky Way is comprised of a central bar, spiral arms, and a halo. The central bar of the Milky Way is about 25,000 light-years in length and is comprised of dense clusters of stars. It is surrounded by four primary spiral arms, where most of the star formation occurs. The halo is a region encompassing the entire galaxy, and it contains dark matter, which astronomers believe makes up 85% of the universe's gravitational mass.

One of the most fascinating aspects of the Milky Way is the supermassive black hole, Sagittarius A*, located at the galactic center. This black hole has a mass of approximately 4 million times that of the Sun and is surrounded by a disk of gas and dust.

Astronomers can study the Milky Way through various methods, such as observing in different wavelengths of light and using computer simulations. They have discovered that our galaxy is constantly changing and evolving, with stars moving around and interacting with each other in complex ways.

The Milky Way's history is also closely tied to that of our solar system. The solar system formed around 4.6 billion years ago from a cloud of gas and dust that was part of the Milky Way's spiral arms. The positioning and movement of our solar system within our galaxy have a profound effect on the conditions that led to the formation and evolution of life on Earth.

Studying the Milky Way is important in helping us understand the universe's larger structure and evolution. It's an ongoing effort by astronomers to observe and understand the fascinating and intricate web of stars, gas, and dust that exist in our galaxy.

BLACK HOLES

Black holes are one of the most mysterious and fascinating objects in the universe. A black hole is a region in space where gravity is so strong that nothing, not even light, can escape its pull. Black holes are formed when massive stars collapse under the force of their own gravity at the end of their lives.

There are two main types of black holes: stellar and supermassive. Stellar black holes are created when a massive star explodes in a supernova and its core collapses in on itself, causing a singularity to form. This singularity is an infinitely dense point where the laws of physics as we know them break down. Supermassive black holes, on the other hand, are found at the centers of galaxies and can be millions or even billions of times the mass of our sun.

One of the most fascinating aspects of black holes is their event horizon, which is the boundary around a black hole beyond which nothing can escape. The event horizon is sometimes referred to as the "point of no return." Anything that crosses the event horizon is pulled inexorably towards the singularity at the center of the black hole.

Black holes are also fascinating because they give us a glimpse into the fundamental nature of space and time. According to Einstein's theory of general relativity, gravity is the result of the curvature of spacetime. Black holes are the ultimate test of this theory, as their immense gravity causes space and time to become infinitely curved around the singularity.

Although we cannot directly observe black holes, we can observe the effects of their gravity on nearby objects, such as stars and gas clouds. We can also study the behavior of matter as it approaches the event horizon, which can give us insights into the basic properties of black holes.

Despite the fact that black holes are amongst the most exotic objects in the universe, they are surprisingly common. There are likely millions, if not billions, of black holes scattered throughout our galaxy, and billions more throughout the universe. As our understanding of black holes grows, we may one day unlock the secrets of these fascinating objects, and perhaps even learn more about the nature of the universe itself.

DARK MATTER AND
DARK ENERGY

For many years, scientists believed that the majority of the universe consisted of the matter that we can visibly see – like stars, planets, and galaxies. However, recent studies have revealed that only a small percentage of the universe is made up of this ordinary matter. The remaining 95% of the known universe is composed of two mysterious entities known as dark matter and dark energy.

Dark matter is a form of matter that does not interact with light, meaning it cannot be seen, heard, or felt. Scientists are not sure what dark matter is made of, but it has been detected through its gravitational effects. In fact, it is believed that dark matter makes up about 27% of the universe – nearly six times more than regular matter. Despite its peculiar properties, dark matter is thought to play an important role in the formation and structure of galaxies. In fact, without the presence of dark matter, galaxies would not have formed efficiently and may have not existed at all.

On the other hand, dark energy is a force that is driving the acceleration of the expansion of the universe. Like dark matter, scientists are not sure what dark energy is made of or how it behaves. However, it has been observed through its effects on the universe's expansion. Interestingly, dark energy seems to be evenly distributed throughout space, causing the universe to continue its rapid expansion.

The study of dark matter and dark energy is still in its early stages, and much is still unknown about these entities. However, recent advancements in technology have allowed scientists to study their effects in greater detail. For example, the Large Hadron Collider, a particle accelerator that is operated by the European

Organization for Nuclear Research (CERN), is one of the machines being used to study dark matter. Additionally, recent missions such as the Dark Energy Survey are attempting to give more insight on the nature of dark energy.

In summary, dark matter and dark energy are two of the most incredible mysteries in the universe. They are both invisible, yet their effects can be seen through the powerful forces of gravity and acceleration. As our understanding of the universe continues to grow, we hope to uncover more about these entities and their crucial roles in the cosmos.

THE BIG BANG THEORY

The Big Bang theory is the most widely accepted explanation of how the universe began. It proposes that the universe originated from a point of infinite density and temperature, about 13.8 billion years ago. This theory is supported by a variety of observations and has become the prevailing model of the universe.

The basic idea behind the Big Bang theory is that the universe began in a state of extreme density and temperature. At this point, all the matter and energy in the universe were concentrated in a very small region called a singularity. The singularity existed before time and space began, so there was no "before" it.

After the singularity, the universe expanded rapidly, in a process known as cosmic inflation. This phase lasted for a fraction of a second but resulted in the universe growing exponentially. During this period, the universe cooled and some of the energy transformed into particles of matter and antimatter.

A few seconds after inflation ended, the universe had cooled enough for protons and neutrons to form in a process known as nucleosynthesis. This gave rise to the building blocks of the universe: hydrogen, helium, and a small amount of lithium.

Over millions of years, these atoms clumped together to form stars and galaxies. The universe continues to expand today, and we observe the expansion through the redshift of light from distant galaxies.

The Big Bang theory is supported by many observations, including the cosmic microwave background radiation. This radiation is a remnant of the early universe, and its precise temperature and pattern of fluctuations match what we would expect from cosmic inflation.

While the Big Bang theory has become the standard model of the universe, there are still many unanswered questions. For example, what is the nature of dark matter and dark energy? How did the first stars and galaxies form? Addressing these questions is still a major focus of research in astrophysics.

Despite the limitations, the Big Bang theory has fundamentally transformed our understanding of the universe. It has provided us with a framework to study the universe, and has opened up new avenues for discovery and exploration.

THE EXPANDING UNIVERSE

In the early twentieth century, astronomers started noticing that galaxies were moving away from us. This phenomenon is known as the redshift. The farther a galaxy is, the faster it appears to be moving away from us. This led to the development of the concept of the expanding universe.

The expanding universe theory suggests that the universe was once very small and intensely hot, and has been expanding ever since. This theory was first proposed by Belgian astronomer Georges Lemaître in 1927. American astronomer Edwin Hubble provided observational evidence to support this theory in 1929. He showed that the farther a galaxy is from us, the more it is moving away from us. Hubble's discovery led to the development of the Hubble constant, which measures the rate of expansion of the universe.

The current model explaining the universe's expansion is called the Big Bang theory. According to this theory, the universe began as a single point, known as a singularity. It was incredibly hot and dense, and all the matter in the universe was concentrated in this one point. Then, around 13.8 billion years ago, the universe started expanding rapidly. As the universe expanded, it became cooler and less dense, and the matter started to spread out.

The Big Bang theory explains many observations about the universe. It explains why galaxies are moving away from each other, the cosmic microwave background radiation, and the abundance of light elements like hydrogen and helium. Scientists have also studied the cosmic microwave background radiation, which is the thermal radiation left over from the Big Bang, and it has shown many features of the universe's expansion over time.

However, there are still many unknowns about the expansion of the universe. Scientists are still trying to understand what caused the Big Bang, what dark matter and dark energy are, and what the ultimate fate of the universe will be. In addition, scientists are still looking for more observations to confirm this theory, which would help scientists to learn even more about the universe's history and why we're here.

In conclusion, the expanding universe theory is a fascinating subject that has led to many new discoveries about the universe. The Big Bang theory, in particular, has revolutionized our understanding of the universe's origins and how it has evolved over time. It remains an active area of research and exploration for scientists, and we can look forward to many more exciting discoveries in the coming years.

SUPERNOVAE AND NEUTRON STARS

Supernovae are some of the most spectacular and powerful events in the universe. They occur when a star runs out of nuclear fuel and collapses, producing a massive explosion that can outshine entire galaxies. These explosions can produce some of the most exotic objects in the universe, including neutron stars and black holes.

A neutron star is a type of celestial object that forms during a supernova explosion. As the star collapses, it becomes extremely dense, with a mass greater than that of the Sun but compressed into a radius of just a few kilometers. This extreme density results in an incredibly strong gravitational field, which can distort the spacetime around it. Neutron stars are incredibly hot and emit a range of radiation, from visible light to X-rays and gamma rays.

Neutron stars are also known for their strong magnetic fields, which can be up to a billion times stronger than that of the Earth. These magnetic fields can produce beams of radiation that sweep through space like a lighthouse, known as pulsars. When these emissions are observed from Earth, they appear as a series of regular pulses, which gave them their name.

In addition to neutron stars, supernovae can also create black holes. Black holes are incredibly dense objects that are so massive that they warp the space around them, preventing anything from escaping their gravitational pull, including light. Unlike a neutron star, a black hole has no surface or structure - it is simply a region of spacetime where gravity is so strong that even light cannot escape.

The study of supernovae and neutron stars has revolutionized our understanding of the universe, from the behavior of matter at extreme densities to the prediction and detection of gravitational waves. New telescopes and observatories, such as the Laser Interferometer Gravitational-Wave Observatory (LIGO), are opening up new avenues for exploring these fascinating objects and unlocking the secrets of the universe.

EXOPLANETS: PLANETS BEYOND OUR SOLAR SYSTEM

For centuries, humans have been fascinated with the possibility of finding life beyond planet Earth. And with advances in technology over the past few decades, this idea has become more of a reality than ever before. Astronomers have discovered thousands of planets beyond our Solar System, known as exoplanets, and this number is continuously increasing.

Exoplanets are planets that orbit stars other than our Sun. They can vary in size, ranging from smaller than Earth to larger than Jupiter. Some may be rocky like Earth, while others may be gaseous like Jupiter. Of particular interest are exoplanets that exist within the habitable zone of their respective stars. The habitable zone is the region around a star where temperatures are moderate enough to allow liquid water to exist on a planet's surface, making it potentially suitable for life.

The discovery of exoplanets can be attributed to a myriad of techniques, including the transit method, radial velocity method, and direct imaging. The transit method involves monitoring for slight dips in the brightness of a star as a planet passes in front of it. The radial velocity method measures the periodic wobbling of a star as it is influenced by the gravitational pull of an orbiting planet. The direct imaging method involves capturing an image of a planet orbiting a star.

Scientists have discovered many exciting and unusual exoplanets in recent years. For example, some exoplanets have been found to orbit binary star systems, meaning they have two suns. Others have been found to be tidally locked, where one side of the planet always faces the star, causing extreme temperature differences between the day and night sides.

Additionally, the discovery of potentially habitable exoplanets has raised the possibility of finding extraterrestrial life. While the search for life beyond Earth remains ongoing, the discovery of these worlds has sparked immense curiosity and paved the way for future discoveries.

Overall, exoplanet research is an essential aspect of astronomy, as it not only provides insight into the origins and evolution of planetary systems but also has the potential to revolutionize our understanding of the universe and its place in the grand scheme of things.

THE SEARCH FOR EXTRATERRESTRIAL LIFE

The search for extraterrestrial life has long captured the imagination of scientists and the general public alike. The possibility of discovering life beyond Earth has sparked numerous missions and initiatives aimed at exploring our solar system and the broader universe. But what exactly are we looking for when we search for extraterrestrial life, and how do we go about finding it?

The search for extraterrestrial life begins with the search for habitable environments. Life as we know it requires certain conditions to survive, including a source of energy, liquid water, and a stable environment. Scientists look for these conditions beyond Earth, focusing on planets, moons, and other celestial bodies within our own solar system and in other solar systems.

One of the most promising methods for identifying habitable environments is through the study of exoplanets. Advances in technology have allowed scientists to detect and study exoplanets in unprecedented detail, and some of these exoplanets have been found to exist in the "habitable zone," the range of distances from a star where the conditions are just right for liquid water to exist on the planet's surface.

Another approach to finding extraterrestrial life is to search for biosignatures, or indicators of life. These can include the presence of certain chemicals in a planet's atmosphere, patterns of light absorption and reflection, or even the way light passes through a planet's atmosphere. The search for biosignatures is a challenging one, as we only have a sample size of one (life on Earth) to compare to.

PHILIPP FRÜHWIRTH

In addition to searching for extraterrestrial life directly, scientists are also exploring the possibility of finding evidence of past life on other planets or moons. This can be done by studying the geology and chemistry of these environments, looking for signs of organic molecules or other compounds that could have been produced by living organisms.

The search for extraterrestrial life is a long-term, ongoing effort that requires collaboration and coordination among scientists and organizations around the world. It is a field that is constantly evolving, with new technologies and discoveries leading to new possibilities and avenues of exploration.

Regardless of whether we do eventually discover extraterrestrial life, the search itself is valuable, as it expands our understanding of our place in the universe and the potential for life beyond our own planet.

ASTROBIOLOGY: THE STUDY OF LIFE IN THE UNIVERSE

Astrobiology, also known as exobiology or xenobiology, is the study of life beyond Earth. It is an interdisciplinary branch of science that combines astronomy, biology, chemistry, geology, and physics to understand the origin, evolution, distribution, and future of life in the universe. Astrobiologists investigate the possibility of extraterrestrial life, including microbial organisms, complex life forms, and intelligent beings. They also seek to understand the environmental conditions that are necessary for life to exist, and the impact of life on planetary and cosmic processes.

Astrobiology emerged as a formal field of study in the 1960s, during the space race between the United States and the Soviet Union. The search for life beyond Earth was fueled by the discovery of organic molecules in meteorites and the exploration of Mars and other planets in the solar system. Since then, numerous missions have been launched to study the potential habitability of other worlds, including the Mars rovers and the Kepler Space Telescope.

One of the biggest questions in astrobiology is whether life exists beyond Earth, and if so, where and in what form. Some scientists believe that life may exist in our own solar system, such as on Mars or one of Jupiter's moons. Others speculate that life may exist on planets orbiting other stars, known as exoplanets. So far, over 4,000 exoplanets have been discovered, and some of them are located in the habitable zone of their host star, where the temperature allows liquid water to exist.

In order to search for life beyond Earth, astrobiologists use a variety of methods and techniques. The most common method is to look for biosignatures, or signs of life, in the atmosphere or surface of a planet or moon. These can include the presence of certain gases, such as oxygen, methane, or nitrogen, which are produced by living organisms. Another technique is to search for complex organic molecules, which can be the building blocks of life, in rocks or soils.

Astrobiology also raises important philosophical and ethical questions. For example, if we were to discover extraterrestrial life, how would that impact our view of ourselves and our place in the universe? What would be our ethical obligations to protect alien life forms, and how would we communicate with them? These are all important questions that astrobiologists and society as a whole will need to grapple with as we continue our search for life beyond Earth.

COSMOLOGY AND THE FATE OF THE UNIVERSE

Cosmology is the branch of astronomy that deals with the overall structure and evolution of the universe. It is the scientific study of the origin, evolution, and eventual fate of the cosmos. In the last century, astronomers have made significant advances in our understanding of the universe, and cosmology has emerged as one of the most exciting and challenging areas of research in astronomy.

One of the most important discoveries in cosmology is the realization that the universe is not static but expanding. This discovery was made in the early 20th century by the American astronomer Edwin Hubble, who noted that distant galaxies were moving away from us at a speed proportional to their distance. This observation led to the development of the Big Bang theory, which suggests that the universe began as a singular point of infinite density and temperature, and has been expanding ever since.

The study of cosmology has also led to the identification of several components of the universe such as dark matter and dark energy. Dark matter is an invisible, non-luminous material that exerts a gravitational force on visible matter. Its presence can be inferred by observing its gravitational effects on visible matter in galaxies and galaxy clusters. Dark energy, on the other hand, is believed to be responsible for the observed acceleration of the universe's expansion.

The question of the ultimate fate of the universe is still a subject of active research and debate among cosmologists. There are several possible scenarios, including the Big Freeze, the Big Crunch, and the Big Rip, each of which depend on the overall structure and composition of the universe.

The Big Freeze scenario suggests that the universe will continue to expand, and as it does, galaxies will become increasingly isolated from one another. Eventually, all matter and energy will become evenly distributed, resulting in a state of maximum entropy, and the universe will become cold, dark, and lifeless.

Alternatively, if the universe's expansion rate slows down or reverses, the universe could undergo a process known as the Big Crunch, in which the universe would eventually collapse under the force of gravity, ultimately creating a Big Bang-like explosion.

Finally, the Big Rip scenario posits that the expansion of the universe will continue to accelerate, eventually becoming so fast that it will tear apart individual galaxies, stars, and even atoms. The end result would be a universe in which there is nothing left but energy.

In conclusion, the study of cosmology has led to some of the most exciting discoveries in modern astronomy, including the Big Bang theory and the concepts of dark matter and dark energy. While the fate of the universe is still a matter of debate among cosmologists, the work being done in this field promises to shed light on some of the most profound questions facing humanity today.

FUTURE OF ASTRONOMY: EXPLORATION AND DISCOVERY.

As we look towards the future of astronomy, we can expect significant breakthroughs in our understanding of the universe. With advancements in technology and the continued exploration of space, there is much to discover and explore.

One of the most significant areas of future study in astronomy will be the search for habitable exoplanets. With the discovery of thousands of exoplanets within the Milky Way galaxy, scientists are searching for planets that could potentially host life. The use of more advanced telescopes and space observatories, such as the upcoming James Webb Space Telescope, will greatly expand our ability to detect and study exoplanets.

Another area of interest and research will be the study of gravitational waves. The recent detection of these waves has opened up a new avenue of cosmic exploration, allowing us to better understand the formation and evolution of massive objects such as black holes and neutron stars. With the continued development of gravitational wave detectors, we can expect even more significant discoveries in this field.

The future of astronomy also involves space exploration, including manned missions to both the Moon and Mars. With renewed interest in returning humans to the Moon and eventually exploring Mars, there will be new opportunities for scientific exploration and discovery. Developing the technology and infrastructure necessary for long-term space missions will be a significant focus for space agencies around the world over the coming decades.

Advancements in technology will also allow us to better explore and understand the universe. The use of machine learning and artificial intelligence in data analysis will help us quickly and accurately process large amounts of data from telescopes and other instruments. Additionally, new technologies such as adaptive optics and interferometry will greatly improve our ability to observe and study the cosmos.

Finally, international cooperation will play a critical role in the future of astronomy. Collaborative efforts among space agencies and scientists from around the world will be necessary to tackle the major scientific questions of our time. The development of new collaborations and partnerships will promote technological advancements and new discoveries.

In conclusion, the future of astronomy is filled with exciting opportunities for exploration and discovery. With significant breakthroughs in technology and international cooperation, we can expect to greatly expand our knowledge of the universe and our place within it.

www.ingramcontent.com/pod-product-compliance
Lightning Source LLC
Chambersburg PA
CBHW071117220526
45467CB00004B/1922